IMAGES OF THE PAST

FARMING

IMAGES OF THE PAST

FARMING

JON AND DIANE SUTHERLAND

REMEMBER WHEN

First published in Great Britain in 2010 by
REMEMBER WHEN
an imprint of
Pen & Sword Books Ltd,
47 Church Street,
Barnsley,
South Yorkshire.
S70 2AS

A CIP record for this book is available from the British Library.

ISBN 978 184468 113 6

Typeset by S L Menzies-Earl

Printed and bound by MPG

Pen & Sword Books Ltd incorporates the Imprints of
Pen & Sword Aviation, Pen & Sword Maritime, Pen & Sword Military, Wharncliffe Local History, Pen & Sword Select, Pen & Sword Military Classics, Leo Cooper, Remember When, Seaforth Publishing and Frontline Publishing.

For a complete list of Pen & Sword titles please contact
Pen & Sword Books Limited
47 Church Street, Barnsley, South Yorkshire, S70 2AS, England
E-mail: enquiries@pen-and-sword.co.uk
Website: www.pen-and-sword.co.uk

Contents

This is a print that was created in London by the artist William Hincks, dated 20 June 1791. It shows a farmer ploughing, another man sowing flax seed and in the background a man driving a team of horses pulling a harrow. To the left there is a woman with children seated beside a tree and encouraging her husband to take a break from the work. This print was dedicated to Lord Lieutenant General and General Governor of Ireland and is supposed to represent a scene near Scaroa in County Down, now in Northern Ireland.

Introduction

From the sixteenth century the traditional organic agricultural methods that were highly labour intensive in Britain were gradually replaced by a new farming system. This agricultural revolution is greatly misunderstood. It is often associated with the selective breeding of livestock and also the restrictions on access to common land plus the introduction of new systems of cropping. Many people know the names of some of these early agricultural pioneers that are credited with bringing in the agricultural revolution to Britain, such as Jethro Tull. Tull was not, as many people believe, the person that invented the seed drill and this piece of equipment was not used until a hundred years after he had written about it in the 1730s.

In England the agricultural revolution happened around 1750. The population of England was about 5.7 million. Prior to this the agricultural industry found it difficult to feed any more people, but something certainly happened, as a hundred years later the population had risen to 16.6 million and agriculture had to move with it. Output grew and new farming systems were used to rotate crops, land was reclaimed, such as the draining of fenlands in eastern England. The old, low intensity farming was now being replaced by high intensity systems growing arable crops. Woodlands were cleared, upland pastures were being used and around thirty per cent of all the agricultural areas were transformed within a century.

During the industrial revolution the mix of crops changed; low yield crops such as rye were replaced by wheat or barley. Land that had been used for grazing was now turned over to crops, with the animals being fed from the fodder crops, particularly clover and turnips. These crops also helped reclaim lowland heaths and rough pasture to create highly productive arable farms. Turnips were incredibly important; they could be sown in rows and the weeds removed while it was growing. In 1700 fallow land accounted for around twenty per cent of all of the arable land in England but by the 1870s it was just four per cent.

Cereals were also highly important; the yields increased by at least twenty per cent in the hundred years from 1700 and then by another staggering fifty per cent in the fifty years from 1800.

Farmers at this time did not know anything about nitrogen or its importance. Livestock manure was routinely ploughed back into the land and the introduction of legumes, such as peas and beans, helped convert nitrogen in the atmosphere to

This is an aerial photograph of ridges and furrows in the Midlands. By looking carefully at the photograph we can see the patterns of the furrows, or ridges, which show how the land was ploughed before enclosures and hedgerows became commonplace. The land was ridged to help with drainage and in this photograph you can clearly see how the ridges were used to cope with the slopes.

nitrates for the soil. Clover also dramatically increased the amount of nitrogen in the soil, which was so vital for the cereal crops.

The new farming systems were very sustainable; the output had increased without endangering agriculture's long-term viability.

After 1750 agricultural workers could produce more food per head. This meant that fewer workers were needed on the land. This, in turn, enabled a proportion of the former farm workers to move to towns and cities and become industrial workers. In effect, the agricultural revolution made the industrial one possible. In fact by 1850, around the same time that the first photographs were being taken, Britain had the smallest proportion of agricultural workers in the world, at just twenty-two per cent. Britain had a big advantage; there was far more animal power available than in many other countries. By the 1820s machinery was coming into use, which improved cutting and threshing grain.

There are no national agricultural statistics before 1866, so it is difficult to be precise about the pace of the change. But what we do know is that by the time the first photographs were being taken of British farms, workers, techniques and machinery, the entire nature of farming had radically changed and it would only go on changing in the years to come.

This is a photographic print on a stereo card, or stereograph. It was taken by an American photographer around 1905 and it shows farmers at the cattle fair in Killarney, which is a town in County Kerry, now in the south-west of the Republic of Ireland.

We have organized the collections of photographs into six distinct groups. The first four cover England, Scotland, Ireland and Wales. These were all taken at the turn of the nineteenth or early twentieth century. In many respects these photographs are the most exciting and evocative, as they often illustrate lost crafts, traditions and techniques. Many of these labour intensive duties routinely carried out by farm workers, or by those living in the countryside, would be swept away within years. Subsistence farming and labour intensive work would no longer be an acceptable way of life for many people.

The fifth set of photographs is from a collection belonging to James Payne. They depict typical farming scenes in the 1930s and primarily focus on gathering in the crops and haymaking. These photographs show the mix of real horse power, as well as traction engines and early tractors.

The final collection is part of one owned by James Payne. The photographs were taken predominantly in southern England in the 1930s. The photographer seems to have toured throughout Buckinghamshire, Hertfordshire and Kent; although there are expeditions to places further a field, including Ludlow, Wendover and Warwick. The photographer had a keen eye for rapidly disappearing images of traditional rural England. Whilst these are not directly photographs of farming in progress, they illustrate aspects of the lives and the environment of rural people in villages and market towns in what was probably the last decade of relative isolation and tranquillity.

Chapter One

English Farming at the Turn of the Nineteenth Century

Across Britain and certainly in England by the 1860s prosperity in agriculture had all but disappeared. Ten years later it entered a period of depression; profits were down and so was employment. For various reasons British trade abroad had collapsed and in the 1870s there were three years of poor springs and rainy summers that produced poor quality cereal crops, mildewed wheat, mouldy hops, cattle disease and sheep rot. Even the pastures were badly affected. Imports of crops had to be stepped up to feed the nation.

In the late 1870s potato crops failed and there was a cattle plague. The United States was enjoying abundant harvests and their imports only served to drive prices even further down. Land rents in England were very high and thousands of tenants on corn growing lands simply could not pay their landlords.

Even as the century reached its last decade the problems had not disappeared. Growing wheat, barley and oats declined; grazing, dairying and growing vegetables, fruit and raising poultry all increased, simply to help make ends meet. Boiled and tinned mutton had begun to make an appearance in the 1880s, as did beef from Argentina and pork from America. By the late 1890s frozen carcasses were also being imported. Cheese and butter imports were high, as was wool from New Zealand.

The government had to do something to try and save British agriculture. But it was incredibly difficult to make ends meet and there were widespread bankruptcies. The number of people being employed on the land dwindled.

There were two cold summers in the early 1890s, a drought in 1893 and a very poor harvest in 1894. A Royal Commission was appointed in 1893 and it found that the best equipped farms were still profitable and that those with very few workers had flourished, particularly in market gardening and fruit farming.

This photograph was taken around the turn of the century in Devon. Here an elderly farmer is following in the wake of his team early in the year. The ground is being opened up by the plough and then crumbled by the frost, later to be raked by the harrow and then smoothed by the roller. Today's Devon is predominantly cattle and sheep, with some dairy, pigs and poultry. General cropping such as can be seen in this photograph has dropped to barely one per cent. Early farming records in areas such as this tended to be kept in small exercise or pocket books, created by men that were sometimes barely literate. Very few of these farming records still exist. At the Devon Record Office in Exeter there are a handful of account books and farming diaries, some of which cover the later nineteenth century.

The great agricultural depression was at its worst by 1895, but as techniques and crops changed the first few years of the new century saw growing prosperity for farmers. People in the countryside could look back to the 1860s and the early 1870s and recognize that they were now doing at least as well as those great days. Even tenant farmers were doing well.

An Isle of Wight dairymaid is hard at work on the churn in this photograph. This is the first of a number of photographs that were taken by Horace W Nicholls (1867-1941). He was an artist photographer who had grown up in Sandown on the Isle of Wight, where his father had a studio. He was apprentice to his father and later, after working for a chemist in Huddersfield, he took a job as a photographer in Chile. By 1889 he was back and setting up his own studio in Windsor before spending a period of time in South Africa. Nicholls was one of the very first photojournalists and although he was probably best known for his photographs of the Second Boer War, he photographed a large number of domestic scenes and portraits. One of his best known photographs is one that he took of soldiers being inspected by King George V at Armoury House in London in September 1914.

At the turn of the century all poultry farming was effectively free range, or at least what is now known as yarding. Raising chickens has always been a typical farmyard activity. Most countryside families kept a flock for eggs, meat and for bi-products such as feathers and manure. Any extra birds were sold at market, or bartered. In 1910 Joel M Foster, the founder of the so-called Million Egg Farm, wrote a book to promote humane and environmentally responsible mass production of eggs and birds. He also designed and sold incubators, sanitizers and feeders.

A farmer poses with his children as he takes a break from his daily chores in this photograph. In the background we can see hay piled up on a wagon, which is then being stacked onto a hayrick. Hay is a mixture of long grass and wild flowers that would need to be dried and used for animal food in winter. Hayricks were once a very common sight in the countryside. The long grass was cut and then dried in the fields before being taken by horse and cart and stacked into a rick. The rick was thatched with wheat or rye straw and was then secured using twine tied to short pegs, which were knocked directly into the hay. Depending on the part of the country, the hayricks were different shapes. Bales of hay were cut from the hayrick the following winter, using a hay knife, to provide winter fodder for the animals. This process was called hay trussing. Ricks effectively became redundant by the mid-twentieth century, as there were more economic ways of storing the hay by this time.

A rather idealized photograph here, of Suffolk girls from Huntingfield working as extra labourers to bring in stooks of oats. A stook is effectively a bale or a truss. This was a term that dated back to at least the middle of the sixteenth century. It is believed that the word is of German origin and may have originally been used to describe heaps or bundles of corn or flax. In other parts of the country they were called cocks, which was a cone-shaped pile of straw or hay. Huntingfield is still a vibrant agricultural village in the Blything district of Suffolk and close to the market town of Halesworth. Thomas Charles Mudd, at the age of sixteen, left Huntingfield in Suffolk in 1912 and boarded the *Titanic* at Southampton, as a second-class passenger, having paid £10 10s for the voyage. Mudd was one of the many unrecovered casualties when the ship sank.

This photograph is believed to have been taken on the Isle of Wight and shows three horses pulling a harvester that is reaping the standing corn. It also binds it into sheaves, as can be seen to the left of the photograph. There is also a substantial hayrick in the background. This appears to be a typical horse drawn reaper and binder, which were common sights across the whole of England. As we will see in the next photograph, where the men are hand reaping with scythes, the horse drawn version would be used for barley, oats and wheat, usually during the month of August. The corn would have to be stacked in the farmyard and then thatched in case of rain. In July each year there would be a short breathing space before the harvest began. The month before would have seen the hay harvest.

The old method of harvesting corn, using scythes, can be seen here. In smaller and less accessible fields scythes would still be used well into the twentieth century. Women would usually follow the reapers, tying up the sheaves and standing them up in bundles to dry. They would normally be left in the field for three Sundays, until they were brought to the farmyard for safe keeping and sale. School did not resume in September until the harvest was over, as even the children were needed. At the end of the harvest a supper would be put on by the farmer. The extra money that had been earned by the farm workers would pay off their rent arrears and buy them a few new clothes and boots in the market towns. By October the beginning of the farming year would start again, with farm tenancies changing hands and workers looking for new jobs.

This incredible photograph was taken on the very edge of Portland Bill. The sheep are taking advantage of every available inch of grazing, even though the terrifyingly choppy English Channel is directly below them. Portland Bill is, of course, in Dorset and particularly known for its three historic lighthouses, the first of which was completed in 1716. Portland Bill and nearby Chesil Beach have always been considerable hazards to the many vessels approaching Portland and Weymouth. Although we cannot tell what type of sheep these are, the most prized in the county are the Dorset Horn sheep. A breeders' association was established in 1891. The Poll Dorset sheep was developed from one of the oldest breeds of sheep, the Dorset Horn. Both types were unique and produced high quality wool. Modern-day Poll Dorset types came from Australia. The county also has Portland sheep, which are small, horned and have brown or tan faces and legs. The lambs are particularly appealing, with foxy brown wool that changes to grey or white, but many still retain red or fawn patches. Nowadays the breed is categorized as being vulnerable.

This is a photograph taken at Bradford Peverell of agricultural workers enjoying a drink outside a tap house after haymaking. Bradford Peverell is near Dorchester in Dorset and still retains its predominantly agricultural existence. It was the birth place of Dorset's most famous historian, John Hutchins (1698–1773). The church there dates from 1850 and has glass dating back to the thirteenth century, taken from New College in Oxford. There has been a longstanding farming tradition in the area and in fact a short distance from the village there is the New Barn Field Centre, which is a living museum recreating an Iron Age homestead, showing the life of the Celts that lived in the region.

A pair of workmen are harvesting seaweed at the ebb tide here, on a beach near to Penzance in Cornwall. The seaweed is then carted back to the farm and used for manure, costing the farmer absolutely nothing beyond the wages for the men and fodder for the horses. Seaweed has quite a connection with farming; in parts of Wales it is used as an ingredient in bread. Agar is also produced from seaweed. In this particular instance, however, it is not designed to be eaten directly, but it still enters the food chain. In Cornwall until relatively recently it was not uncommon to see tractors and trailers collecting seaweed from the beach. Seaweed was gathered all along the Cornish coast and at the turn of the century it was not unusual to find advertisements in the newspapers offering cartloads of seaweed for manure. The seaweed around Penzance was particularly abundant when there was a north wind and farmers were lucky to get two loads, such was the demand. They would pile it onto the fields, or mix it with manure, known as dressing.

This is a second photograph from Cornwall, showing the deep valleys that run down from the moorland to the sea. These are sheltered, well wooded and lush due to the streams that run down them and, as a result, they make excellent arable and pastureland. It does, however, make the life of a farmer far more difficult and strenuous, although the horse seems to be enjoying himself pulling along a small roller, whilst a second tethered horse is exercised in the background. The vast majority of the farms in Cornwall were unsuitable for modern machinery and horses or oxen were still used well after they had been replaced by tractors in other parts of the country.

This is an ingenious labour saving method of hay-stacking on a Kentish farm. High masts have been erected, with attached pulleys, to which are fastened a large cage fork. The pulleys are worked by the horse in the foreground and the hay is being raised in the fork from the cart on the right, to where the men are building the haystack on the left. Although this system looks very complicated and cumbersome it is saving the men a great deal of time and effort. Hayricks were a major target in an uprising by rural workers in Kent in the 1830s. They had seen their wages drop and the new threshing machines were threatening their livelihoods. The supposed leader of the riots was called Captain Swing. The first threshing machine was destroyed on August 28 1830 and by the end of October over 100 threshing machines had been destroyed in east Kent alone.

A peaceful, rural scene is shown here. This photograph was taken in a quiet Herefordshire lane at the early part of the twentieth century. Herefordshire was well known for its cattle, sheep, hops and cider. Here three horses are pulling a cart that has a large trunk of wood onboard. The major cereal of Herefordshire was wheat, being sewn in the first half of October. It was harvested at the beginning of August and into the first week of September the following year. In the early years the majority of the wheat was cut by Welsh farm workers. In the following year, between February and mid-March the farmers would sew beans or peas on the same ground and then it would be left fallow for a year. Farm workers tended to be hired in May, although day labourers would also be used in busy periods and part of their pay would be a gallon of cider a day.

Unfortunately we do not know where this photograph was taken, but it charmingly depicts a grandfather in his hobnail boots, sitting patiently on the back step of his cottage, whilst his granddaughter covers his eyes. By the 1901 census the population of Great Britain had risen to just over 38 million; it was growing at a rate of around one per cent per year. At that time the average number of children per family was 3.5. Life expectancy was also better, with the population over fifty standing at around fifteen per cent. Life expectancy for men was forty-five years and for women forty-nine years. There would not be significant changes to this until the 1930s.

This is the first of two photographs that were taken in Worcestershire at the turn of the century. The first is of two older men that have clearly worked for the majority of their lives in the fields and have survived to a ripe old age on a diet of bread and cheese, washed down by beer from a stone jug. It is believed that the man on the right, leaning on his sticks, was the last person in Worcestershire to wear the smock, or slop, which was once a distinctive feature of rural dress. By the late nineteenth century this kind of smock had evolved into a child's garment, but during the nineteenth century it had been widely worn by rural workers. It was also worn by factory workers, but this was less common. Some Morris dancers today still wear this type of smock. Unfortunately there are very few photographs of workers in this type of clothing, as amateur photography was virtually unknown until 1900.

This old farm worker is enjoying a well earned lunch of cheese and bread and has a handy stone jug of beer or cider beside him. He may well have kept this jug cool inside the hay stook. Jugs such as these can trace their ancestry all the way back to the Celts, as they have been found in a number of graves and sites well before the arrival of the Romans. Woe betides any farmer who did not have stone jugs of beer or cider sitting in the shade at the edge of a field to quench the thirst of the workers. Traditional cider making counties included Herefordshire and Somerset, although it was not unknown for cider to be produced wherever crops of apples were being grown. Somerset for many years has had the reputation of being the cider making centre of Britain, with ideal conditions for growing apples, particularly around Wedmore, Glastonbury and Martock.

The village blacksmith, who for centuries was at the very heart of rural life. Although we do not know exactly where this blacksmith was working, his importance could never be overestimated. Here we see a horse standing patiently whilst the blacksmith completes his task of hammering in the last few nails to the shoe. The blacksmith would be incredibly important; he would not only make horseshoes, but also tools, agricultural implements, cooking utensils, gates, railings, lighting fixtures and in times of war, weapons. The term blacksmith relates to the fact that iron was considered the black metal due to the layers of oxides that form on the surface when it is heated. Arguably, the term smith comes from the word smite, which simply means hitting. Logically then a blacksmith can be described as an individual that hits black metal. As machines in factories took away many of the traditional crafts that had been performed by blacksmiths, many of them became farriers or shoers of horses.

This is a photograph that was probably taken in Lincolnshire in the early part of the twentieth century. It shows three horses pulling a machine digger to help harvest the potato crop. In England and Wales by 1921 some 558,000 acres were devoted to the growing of potatoes and the total yield was around 3 million tons. The most famous part of Lincolnshire, in the south-east and most closely associated with potatoes was Holland, with the town and small port of Boston at its heart. The land, much of which had been reclaimed from the sea and was therefore rich silt soil, was perfect for growing potatoes. In 1929 the average yield per acre over a ten year average was 250 bushels, or 6.7 tons. It was only beaten by the county of Norfolk, averaging 370 bushels per acre.

Another stage in harvesting the potato crop is shown here. This is a potato raiser in action. It is being drawn by three horses and is the last machine used in the process, starting with potato planting machines and then potato sprayers. Men follow the machine, undertaking a particularly backbreaking task of gathering up the potatoes into baskets, which the mechanism of the potato raiser has excavated without damaging the tubers being lifted from the ground. In a newspaper advertisement in December 1881 an impressive selection of potato ploughs and potato raisers were being advertised. The potato plough, using a pair of horses, would raise three or four acres of potatoes a day. It was guaranteed to leave fewer in the ground than conventional digging. The roots were also raised with less bruising and scratching. The potato raiser in this photograph resembles a Howard's machine, which was of simple construction and easily managed. The potatoes were spread to the left or right of the implement as it went along, which meant that the field could be worked continually from side to side without having to wait for the gatherers to catch up.

The next stage of the potato harvest shows the crop being poured into sacks from the gathering baskets. This must have been extremely hard work, with the farm workers having constantly to stoop to pick up the tubers and then pour them into sacks, which then had to be carried into the carts so that they could be taken back to the farm for storage. It must have been heartbreaking if the potatoes had caught a disease like blight. Potatoes were particularly vulnerable to bacterial, fungal and viral diseases. They were also often prey to parasites. Even if they were diseased they would still have to be taken out of the ground and the soil treated and left fallow in the hope that any bacteria or spores would die off in the winter months. The worst examples of potato diseases took place in both Ireland and Scotland in the middle of the nineteenth century. The Irish Great Famine is perhaps the most famous. As a result of the blight at least a million people died and another million emigrated. Blight struck the Scottish highlands in the 1840s, again causing hundreds of thousands of people to emigrate. Most of the crofters were very dependent on potatoes as their main food source.

This photograph shows the workers making a clamp. This was a typical way in which to protect the potatoes on the land throughout the winter, without running the risk of damage from the frost or damp. The men would dig a trench alongside a hedge and would then load the trench with the potatoes before covering them with a thick layer of thatch. On top of the thatch would be placed a layer of earth. In theory the potatoes would be safe until the following spring. At this point some of them could be eaten, whilst those that were sprouting could be used as seed potatoes.

A young girl strips the hops off the bine in Kent in this photograph. By the beginning of the twentieth century around a quarter of a million people travelled to Kent to work as hop pickers each summer for the annual harvest. The hops were grown on flexible branches known as bines. They were trained to grow along strings and wires attached to poles. Workers would often have to operate on stilts attached to the strings to reach the highest wires. In the spring the bines had to be 'twiddled' to encourage them to grow on the strings. The hop picking harvest began in late August and went through into September, with picking beginning at dawn. This girl has one of the typical large baskets. She has pulled down the bine, laid it on the basket and is picking off the hops, or the flowers of the plant.

Hop pickers are pouring the hop flowers from a basket into a sack here. The pickers had to be very careful about not dropping leaves into the baskets. The hop sacks were weighed by a tallyman and calculated by the bushel (around 35 litres). Pickers would be paid per bushel. After lunch children played in the fields, while their parents continued working. About a third of the seasonal workers came from the east end of London, providing them with valuable additional income. Mothers and children would spend the school holidays in Kent, being joined by the husband at the weekend. Sometimes families would move down to Kent for the whole of the summer, living in makeshift huts made of corrugated iron. They cooked over open fires and collected water from local pumps or wells.

This photograph shows a tallyman holding one of the long sacks, or pokes. A poke would hold around ten bushels. The man on the right is a booker, who is recording the amount that has been picked. After the hops were weighed they were carried off to kilns or oast-houses, where they were dried. Around ten hours were needed for the drying process and after they had been cooled they would be packed, ready to be sent to the brewers. A large proportion of the seasonal workers were travellers, who would gravitate towards Kent for the summer. In many respects they would be far more prepared to cope with the alfresco living conditions, either by using their own horse drawn caravans or by erecting tents. Some would go to Kent not just to pick hops but also to help the fruit growers. By the 1930s the heyday of the hop pickers was over and much of the work had been taken over by machines. People were also being given paid holidays, which meant that hop picking holidays became less popular. By the 1960s the annual migration from the east end had virtually disappeared and the train companies no longer ran their hoppers' specials.

Butter making, using a churn, in Broseley, Shropshire can be seen here. Broseley is close to Ironbridge. Shropshire used to be very heavily forested. It was seriously depleted when the timber was being used as a local building material and many of the early cottages were constructed by carpenters rather than stone masons or bricklayers. Deforestation became a major concern, as Ironbridge is closely linked with the industrial revolution and whilst huge amounts of coal were used in iron manufacture, the impact on local forests was also severe. The famous Ironbridge itself was built in 1779. It linked Broseley with Coalbrookdale and it actually led to the foundation of the town of Ironbridge, which is now a World Heritage site. The butter churn was used to transform cream into butter, using a totally mechanical process, with the milkmaid using a crank to rotate a device inside the churn. The agitation disrupts the milk fat, breaking down the membranes surrounding the fat. The clumps become known as butter grains. Air bubbles are introduced via the mechanical churning and the butter grains become denser. This produces buttermilk. The fat globules eventually form solid butter and separate from the buttermilk, which is then drained off. It is believed that the word butter comes from the Greek for cow cheese.

Another photograph from turn of the century Shropshire, this time showing hand shearing on a farm, to the obvious fascination of an assembled group of children. During the thirteenth and fourteenth centuries wool from Shropshire was considered, arguably, to be the best in the world. It was an incredibly important source of revenue for the church. By the beginning of the twentieth century Shropshire sheep tended to be old native stock crossed with Southdown sheep. So successful were the Shropshire breed that they were exported to the United States, particularly Virginia, in the 1850s. More were sent to Maryland in the 1860s and after that literally thousands made their way across the Atlantic and most states in America had flocks of Shropshire breed, as did several areas of Canada.

An old village thatcher, with an iron tooth rake, paring knife, bell-hook and forked stick carries off a load of straw from a field in this photograph. The thatcher would use tarred cord and wooden pegs to secure each course of thatch. He would start at the bottom of the eaves and work upwards. Whilst reed was the best type of thatch, it often proved to be too expensive, which meant that straw from wheat or rye would tend to be the most common thatching material.

This type of man would have been a typical sight across the length and breadth of England. His jacket appears to be army surplus and from its cut the photograph probably dates to the beginning of the twentieth century. This type of man was the blueprint for the Jack of all Trades. His small, multi-purpose cart carried all of his travelling equipment and a rudimentary tent shows that he was well used to living in rough conditions wherever his work might take him. Note that the man only has one arm, which may suggest that he is an ex-soldier. He travelled around the country in an open cart, looking for casual work on farms and in villages. This kind of individual would be a rat catcher one day, a thatcher another, or a simple farm labourer. During the summer months he would live in his portable shelter and hopefully retreat to some more permanent accommodation when the winter months set in.

A shepherd is driving his flock through the shady lanes of the village of Fittleworth, near Chichester, now in West Sussex, in this photograph. Fittleworth at the turn of the century was well known for game fishing. Sheep rearing was very important in the area and a large number of flocks enjoyed the grassy slopes of the South Downs. Fittleworth has two bridges, both of which were made of stone. About a mile and a half to the south-east is Coates Castle. The village dates back to at least the twelfth century and from the middle of the sixteenth century the Swan Inn operated as a coaching inn and a place where royal couriers would change their horses. The village itself is on the River Rother, near Pulborough. In the early 1900s it was the home of the artist Charles Sims. The village was also the former home of the composer Edward Elgar, who lived in a thatched cottage called Brinkwells.

This is the first of three photographs that were taken in Burton Bradstock in the late nineteenth century. This particular photograph shows threshing at Burton Manor Farm. In the 1880s fifty-six of the men living in Burton Bradstock were agricultural labourers, twenty-one were general labourers, and twenty-six were flax mill hands. The majority of the women worked as net braiders. Burton Bradstock was typical of many of these small villages. From 1830 to 1870 the population had hovered around 1,000 but by the 1880s it had dropped to below that number for the first time in nearly seventy years. It would continue to drop for the next 100 years. The buildings in Burton Bradstock are nearly all constructed from the abundant stones taken from the nearby cliffs.

Haymaking in Burton Bradstock, probably in the early 1920s, can be seen here. At its height there were at least ten working farms in the parish, all of which produced milk and provided a reasonable income for the owners. In fact the dairy herds were as important as the flax mill as the centre of village life. Wheat, barley and flax were grown in the parish and the larger farms attracted seasonal workers. The wheat and barley were mainly grown for animal feed. Modern-day visitors can see the old farming methods at the Bredy Farm Museum in the village.

This photograph was taken on Burton Manor Farm, probably in 1930 and shows a three-horse team being used to harvest barley. Barley would have been used for animal feed and could also be used for malting, as it was an ingredient in both beer and whiskey. In English folklore John Barleycorn was seen as the personification of barley. In the song associated with John Barleycorn he suffers attacks, death and other indignities, which match the stages of barley cultivation. In the popular hymn *We Plough the Fields and Scatter* he is also mentioned around the time of the harvest festival.

This woman is collecting kelp from one of the sea lochs on the coast of Skye. In the nineteenth century this was always done by hand and invariably people would walk back to the farms, or crofts, as even the furthest loch was fewer than five miles from the croft. There was an enormous abundance of seaweed most of the year. The kelp industry itself had collapsed in the 1820s and many of the inhabitants of Skye and the Highlands were evicted from the best farming lands and had to settle on the barren coastal strips. The landless peoples, of which this woman is clearly one, were relocated to the coast to scratch a living in what remained of the kelp industry. Kelp had been used to produce soap and other products. It had to be burned, often in stone-lined trenches or kilns, with the fire being kept going with dried heather and hay. This would produce soda and potash, which would be used not only for soap making but in the glass industry and also for bleaching linen.

Chapter Two

Scottish Farming at the Turn of the Nineteenth Century

For generations a fundamental system of farming had been used in Scotland, known as the runrig system. These were elongated s-shaped plots that were divided between the local communities in the intensely cultivated area closest to the villages, known as the infield. Beyond this was the outfield, which was less intensively farmed and was generally used to graze livestock. In effect, around each village was a tiny amount of cultivation, surrounded by largely infertile areas that were never ploughed or used for grazing.

Land management changed dramatically in the seventeenth century when the old system was swept away and tenant farmers would be required to pay rent in cash. New ways of farming were developed; crop rotation was used, bigger expanses of land were brought into production, peat bogs were cleared and marshes drained.

Productivity increased, but the landlords squeezed every penny out of their tenants.

In the eighteenth century lowland Scots were the first to feel the negative effects of the new system and, overwhelmed by the agricultural revolution, many left the countryside to head for the towns and cities, or to emigrate to America. After 1745 and the defeat of the Jacobite Rebellion the clan system was destroyed and the Highlands were becoming commercialized. By the beginning of the nineteenth century, in the eyes of the landlords, sheep were far more profitable than people. Thus began the Highland Clearances.

Many of the people were living in subsistence conditions, on unproductive land. They had to work for their landlords in order to survive. By the 1840s the kelp industry had collapsed, which meant the landlords needed fewer labourers, and then there was the potato famine. Thousands fled Scotland, bound for Australia, America, Canada and New Zealand, whilst others drifted into the industrial cities of Scotland and England.

In certain parts of Scotland the old ways continued, particularly on the islands. The crofters retained their way of life, scratching a living from the land and from the sea. The photographs in this collection capture that last generation that lived a hard and difficult life, harking back to the old ways of Scottish rural living.

Nowadays some seventy-five per cent of Scotland is used for agriculture and including whisky exports the industry is worth £2.4 billion. Around eight per cent of the population works on the land: about 65,000 people.

Sheep farming was incredibly important in the nineteenth century. Sheep numbers in the Highlands and the north of Scotland, for example, increased by over twenty per cent between 1855 and 1880. After 1895 there was a decline in the Highlands, but an increase in the lowlands. In the first decade of the twentieth century there were 7.1 million sheep in Scotland. However, since the Second World War there have been further increases. Commercial sheep farming in its various guises can be traced back at least 250 years. The stocking rate was around 1.37 sheep per hectare in the late nineteenth century.

In nineteenth-century Scotland farming was by far the largest industry. The agricultural revolution brought greater diversity. Regions began to specialize: there were beef-rearing crofts in the north-east, arable farming in the south-east and dairy farming in the south-west.

The farmsteads themselves provided shelter for the livestock and horses, they were used to process crops and animal produce, as well as store food, and they would also act as a centre for farming implements and the storage of manure. Most of the livestock buildings were for cattle. In these byres the cattle would be tethered in a line. Cows in dairy byres were housed and milked in the same place. Beef cattle were overwintered in the byres.

When threshing machines began to make an appearance during the late eighteenth and early nineteenth centuries new two-storey byres were constructed to hold them. Most of the power came from horses, waterwheels and sometimes windmills. Later they would be powered by steam engines.

Arable farm districts often had their own dairies, where cheese and butter could be made, but it was central, west and south-west Scotland that concentrated on commercial dairy farming.

Many of the buildings of the nineteenth century used local stone and had either slated or pantiled roofs, although this depended on the district and the purpose for which the building had been constructed.

Crofters on the island of Skye, using an ancient millstone to grind corn, are featured in this photograph. It was written: 'In Skye one is free of one's century; the present wheels away into silence and remoteness.' Crofters would tend to work from early June until the end of August on the kelp harvest, simply to subsidise, and in some cases to replace, their traditional agricultural labour and animal husbandry. The population of Skye had peaked in the 1840s at just over 23,000. By the 1890s it had dropped to under 16,000 and it continued to fall until the 1980s. Crofting still remains important and even today there are around 2,000 crofts on Skye. But only around 100 of these are big enough for the crofter to earn a livelihood from the land.

Farmers on Skye are preparing their land for the potato crop here. Although the crofters on Skye at this time were mainly concerned with sheep, cattle, fishing and distilling, the moors and hill lands were used for pasture and the growing of potatoes and turnips. The phrase 'to be fired' is said to have originated either from Skye or the Highlands, dating back to the Highland Clearances. A landlord would literally burn down a crofter's home to put the family out of work. The largest town on Skye is Portree, with a scenic harbour surrounded by hills, and it boasts brightly painted houses. As part of the Inner Hebrides, around half of the population still speaks Gaelic and many people worship with the Free Church of Scotland, so the Sabbath is still strictly observed. The most popular sport is shinty. Skye is the largest island in Scotland after Lewis with Harris. Its coastline is effectively a series of peninsulas and has some of the most dramatic, mountainous terrain in Scotland.

Kelp burning on the Orkney Islands can be seen in this photograph. There are almost seventy islands or islets, of which, at the turn of the century, around thirty were inhabited. This photograph was taken at Birsay, a small northerly village on Pomona. The kelp industry required hours of backbreaking work – the kelp had to be gathered, dried, fired and then carted. It would need twenty tons to produce one ton of saleable product. The kelp would be burned from between four and eight hours. As in this photograph, women would often watch the fire and make sure that it was burning steadily, which was a considerable skill. Once the fire was going well men with iron mallets or hooks known as kelp irons would pound the weed into a mass. It would then be covered with stones or turf and left overnight. The kelp ash would be cool enough the following morning to be broken up for transportation by boat.

A Highland shepherd, with his inseparable companion, can be seen here. These highly trained and hardy animals were as gentle with the lambs as the shepherd. Writing in 1865, John George Wood, in his *Illustrated Natural History*, described the Scottish sheepdog, or collie, as being:

> Sharp of nose, bright and mild of eye, and most sagacious of aspect. Its body is heavily covered with long and woolly hair, which stands boldly out from its body, and forms a most effectual screen against the heat of the blazing sun, or the cold, sleety blasts of the winter winds. The tail is exceedingly bushy, and curves upwards towards the end, so as to carry the long hairs free from the ground. The colour of the fur is always dark, and it sometimes variegated with a very little white. The most approved tint is black and tan; but it sometimes happens that the entire coat is one of these colours, and in that case the dog is not so highly valued.

A crofter on the Isle of Skye is shown in this photograph. It was not until 1886, when an act was passed to provide security of tenure for crofters in Scotland, that tenants were encouraged to improve the ground under their control, in the hope that it could be passed down the generations. There are estimated to be around 13,000 crofters in the Highlands and islands of Scotland today. Taken together with family members this amounts to around ten per cent of the current population of the area. The most concentrated areas are in the Shetlands, the Western Isles and Skye. Crofting has its roots in the eighteenth century, when landowners parcelled up their land in return for an annual rent.

A shepherd is shown here with his flock near Strathyre, in Perthshire. The valley was first settled by crofters who cleared the land to make room for their sheep. The name Strathyre comes from the Gaelic for broad, winding valley. For generations it had been used as a highway by drovers, bringing their herds of Highland cattle to market. It is also the location of the fictional fugitives David Balfour and Alan Breck in Robert Louis Stephenson's *Kidnapped*, set in 1751. The way of life in Strathyre changed in the 1870s with the arrival of the Callander and Oban railway. The line snaked around the west side of Loch Lubnaig and then crossed the river before running through the valley. This made the area a popular resort, being at the heart of the Trossachs. The area is also associated with Robert MacGregor, or Rob Roy. He was a notorious cattle and sheep rustler who became a legendary character in his struggles against the Duke of Montrose in the early eighteenth century. He is in many respects a Scottish Robin Hood.

Shetland women and their ponies are heading for Lerwi market in this photograph. The women are knitting as th head along the road towards the port. In the panniers c the ponies are butter, eggs and bottles of milk, ready for sale at the market. This photograph was probably taken around 1900. Lerwick can trace its history back around 3,000 years. It was an important fishing centre and benefited from the North Sea oil boom in the 1970s. Shetland ponies have been kept on the island since the Bronze Age and they are known for their heavy coats, short legs and intelligence. With an official maximum hei of forty-two inches at the withers, they are willing to be used for riding, driving and for carrying packs.

A Perthshire farmer is working on his land here. He would be predominantly involved in tending sheep, as well as growing a range of crops. Perthshire sits in central Scotland and had its own local government until 1930. It has always been known as the 'big county', with its high mountains and its vast agricultural lands in the east. Samuel Lewis, writing in 1846, said of Perthshire:

> The surface is remarkably varied, and comprehends an extensive highland and lowland district; the former, to the north and northwest, constituting a considerable portion of the Grampian range; and the latter, which is the more extensive, lying to the south and southeast. It abounds with the richest scenery of every variety; is beautifully diversified with mountains and valleys, wide and fertile plains in the highest state of cultivation, rising grounds, and gentle undulations; and is enlivened with numerous streams and picturesque lakes.

This photograph shows a sheep farmer in Peeblesshire. The county of Peeblesshire disappeared after a local government reorganization in 1975. The area became known as Tweeddale. The main town was Peebles and although Peeblesshire has disappeared the name is still in common usage and is preferred to Tweeddale, or Scottish Borders. In the book *Imperial Gazetteer of Scotland* by John Marius Wilson, he described in 1868 Peeblesshire as being:

> An inland county in the southern division of Scotland. It is bounded on the north and northeast by Edinburghshire; on the east and southeast by Selkirkshire; on the south by Dumfrieshire and on the southwest and west by Lanarkshire.

The population of the county at the turn of the century was around 15,000. It had had a fairly stable population for some years, increasing if anything from the early nineteenth century. Peebles was very important in the wool industry well into the twentieth century. It had become a royal borough in 1152, during the rule of King David I.

A shepherd and his flock are shown here in winter in Lanarkshire. Even in the lowlands of Scotland the climate was always particularly harsh in winter and although sheep breeding was a profitable occupation, it demanded a great deal from the shepherds and from their faithful dogs. The shepherd's year would begin in August and September, when ewes would be prepared for mating. The shepherd would have to make sure that they were in good condition and physically fit. In October the ewes would be mated with a ram. The date of the mating was carefully timed so that the lambs would be born when the grass was at its lushest. During November, December and January the sheep's diet would have to be supplemented with hay, silage and root crops, such as turnips. Lambing would begin in February and go through into March. Until April additional food would have to be available for the ewes and the lambs in case there was insufficient grazing. The sheep were at their highest risk in May, from worms and fly attacks. This would mean dipping. The sheep would be shorn in June, with an average shearer managing to shear around eighty ewes a day, but experts managing around 300. By July the lambs would have been weaned and separated from their mothers. The majority of them would then be prepared for slaughter, although the best lambs would be held back for future breeding.

Shepherds in the Highlands drive their sheep along a deserted track, heading for home. One of the most commonly bred sheep was the Scottish Blackface, which was more than capable of coping with exposed areas like the Scottish Highlands. The animal is also often referred to as the Scottish Mountain or Highland sheep, or the Scotch Horn. They originate from the Anglo-Scottish Border region, probably as far back as the twelfth century. In 1503 King James IV of Scotland established a flock of 5,000 of them to the south of Peebles. Around thirty per cent of all sheep in Britain are now Scottish Blackface. They have long, coarse wool, which protects them from moisture and from the wind. They are capable of surviving in the harshest of climates. They can now be found across the whole of Britain, with the most common variety being the Perth, which has a larger frame and a longer coat. They were shipped out to the United States in 1861, and were introduced to Northern Ireland. The best Scottish Blackface wool is used to make Harris Tweed.

Here we can see the harvesting of wheat in a Highland glen. Even though the majority of the best land had been taken for sheep rearing by landowners, pockets of decent arable land still provided vital space for growing crops, even for the crofters. Many of the Highland families had left the Highlands by the end of the nineteenth century, but many of those that survived had to be basically self-sufficient. The largest arable farms today are in the Lothian and Border regions of Scotland.

Chapter Three

Irish Farming in the Late Nineteenth Century

This small collection of photographs was taken around fifty or sixty years after the Great Famine in Ireland, during which a million people died. This was caused by a blight that devastated the potato crops throughout Europe in the 1840s. Around a third of the Irish population were dependent on potato as their main food source. It was to prove to be the catalyst that would see at least a million people emigrate.

Evidence has been discovered in Ireland that farming has been going on for at least 6,000 years. Various enclosures have been found and the landscape of fields, with hedges and walls, which we see today dates back to the eighteenth and nineteenth centuries. Irish historians often refer to this as the age of improvement, when technology was being used alongside science to create new implements, farming systems and livestock breeds. Irish farming history is dominated by the century-long tensions between rich landowners and tenant farmers, each blaming one another for the problems that the agricultural industry has suffered over generations.

For many centuries Irish farmers relied far more on livestock than on crop production. There was a major shift toward arable farming in the middle of the eighteenth century, mainly to cater for the enormous grain demands in England during the Napoleonic wars. This gave Ireland the reputation of being the granary of Britain.

A so-called land war in Ireland raged between 1879 and 1903, once again pitting landlords against tenants. There were widespread evictions and rural violence. Certainly during the early nineteenth century cultivation had declined in Ireland and livestock farming was fairly dominant.

Ireland is well known for its own distinctive horses and ponies, some of which are now extinct. Irish farm horses were developed and Irish draught horses became a

Irish men and women are burning seaweed on the shore of the Aran Islands in this photograph. As we have already seen, the ash provided soda salts, which were used in soap and glass making, as well as salts of potash and iodine. Eventually cheaper processes were discovered and kelp burning began to die out. The Aran Islands lie off County Galway, in the Atlantic Ocean, and consist of, from largest to smallest, Inishmore, Inishmaan and Inisheer. William O'Malley, who represented the constituency of Galway Connemara between 1895 and 1918, asked the Chief Secretary to the Lord Lieutenant of Ireland about the plight of the kelp industry in Ireland in February 1901. He was concerned that the industry was being squeezed out of existence and that those working in the industry faced an uncertain future. George Wyndham, then the Chief Secretary for Ireland, promised help and training to teach the islanders better ways of preparing the kelp, but recognized that products such as iodine being brought in from South America would mean that the industry was almost certainly doomed.

recognized breed. Connemara ponies were also recognized as a distinctive breed in the nineteenth century.

Kerry cattle had been recognized as a breed in itself in the 1770s. They were good for both meat and milk. Galway sheep and the Roscommon breed were amongst the most successful lowland animals.

Haymaking was not that common in Ireland before the middle of the nineteenth century, but by the time these photographs were taken it was the biggest crop, providing fodder for livestock. Other key agricultural-based activities included kelp gathering and firing, and of course the collection and use of peat from the wetland bogs and moors.

It is important to remember that the photographs in this small collection relate to a time when there was a united Ireland under British rule, and that the partition of Ireland did not come into existence until December 1922.

Two kelp burners from the Aran Islands are standing next to a stack of drying seaweed here. It was the practice in Ireland to allow the seaweed to dry in the sun for several months in kelp stacks. In the *Meriden Daily Republican* newspaper dated 27 December 1897 a reporter wrote about the typical process of kelp stacking and the life of kelpers:

> Here and there on the island you may chance across a little circle of stone in a retired nook. You need not puzzle long over it as if it were some archaeological find, for it is simply the foundation of a kelp stack that has been piled up inside it, and the calcified appearance of the stones, together with the ashes around them, will immediately reveal its use to you. Far away along the windings of some distant cove you will often see the smoke of these kelp fires rolling heavily out to sea, and the kelp stacks themselves are among the common objects of the seashore, with the long trailing kelp weed drooping down their sides. Kelp is, indeed, the main source of their wealth, and many a voyage has to be taken before a ton of kelp is procured. The weed burns down to a viscous, gelatinous mass under the action of the fire and then cools down to hard and rugged blocks.The kelpers live on fish, milk, tea, bread and eggs chiefly, but seldom on butcher's meat and bacon.

Two boys, described as gossoons, can be seen in this photograph. The word gossoon is believed to derive from the French for boy, *garçon*. At around the turn of the century there was an estimated three million acres of bog land in Ireland. These boys are loading panniers fixed to donkeys with peat. Peat consists of the vegetation of aquatic plants that have decomposed and sunk to the bottom of water-filled depressions. Over the years they become compressed and carbonized. Workmen would remove the sods layer by layer, with long, narrow, sharp blades, known as slanes. The peat sods would then be stacked and allowed to dry for at least ten days. In today's Irish Republic a state-owned company is responsible for managing peat production. Milled peat is used in power stations. Peat briquettes – oblong bars of densely compressed and dried peat – are sold for domestic heating. They are largely smokeless.

A man leads his horse and cart, stacked with cut peat, home to Barna on the coast of Galway. Barna lies to the west of the city of Galway and over the last couple of decades it has effectively become one of Galway's suburbs. At the turn of the nineteenth century it was a predominantly Irish-speaking village, some four miles from Galway. It was set on a sandy road that led by the sea to a crossroads where there were two public houses facing one another. Inland there was a narrow, sandy road, which led over a boggy mountain. On the coastline there was a small pier that had been built to bring in supplies for starving people during the Great Famine of the 1840s. There was a girls' school, a small number of cottages and a post office. The two pubs also served as shops. A typical house would have a kitchen and a parlour downstairs and two or three bedrooms above. There were no bathrooms or toilets. There were usually two huts in the garden with dry lavatories inside.

Two Irish women can be seen here, nearly bent double with the weight of their heavily filled baskets, full of peat. These two women are Irish peasants, mother and daughter. Although peat was clean to handle, cheap and plentiful, it required great strength and resilience continually to replenish the store of heating and cooking fuel at home. The peat would throw out a red and lurid flame and would also fill the room with an aromatic fragrance all of its own, which was much loved and continues to be loved by those that use peat for their fuel. Poverty was particularly rife across most of rural Ireland. Various alleviation attempts had been made, notably by church-based charities, but they seemed to be more interested in forcing religion on the recipients of their charity. Many of the houses were thatched, which was a major financial concern, as the straw and the thatcher had to be paid for unless a local man would do the work in return for labour. A leaking roof in the Irish climate was a disaster. Many of the houses were built of mortared stone and had rough, concrete floors. They would have been incredibly damp had it not been for the peat fires that were always kept burning. The majority of the people harvested the peat themselves and there were always long stacks of it against walls close to every house. Many of the houses had kitchens in the middle of the house and bedrooms at either side to ensure that the heat from the peat penetrated the whole of the building.

A scene on the quay at Inishmaan is shown here. These men have trussed up six pigs ready to be transported across the water to the market of Galway. The pigs would be transported in curraghs, which are characteristic Aran Island canoes. Versions of these had been made for around 1,000 years. By the end of the nineteenth century these canoes would have been made with stout canvas sheeting that had been saturated with tar. They were light and fragile, but very seaworthy, and the Aran islanders were so skilful at handling them that they would put to sea even when steamers operating out of Galway would be lashed to the quayside because of bad weather. The Galway steamers would have called at the islands some three times a week. These pigs would have been restrained as they made particularly bad canoeing companions, and there was always the danger that their hooves would split the eighth-of-an-inch-thick canvas, causing the canoe to spring a leak.

Chapter Four

Welsh Farming in the Late Nineteenth Century

By 1900 around twelve per cent of the population of Wales was employed in agriculture. In certain areas, such as in Cardiganshire, it was as high as forty per cent. Here, for example, most of the farms were mixed, keeping animals and growing crops. Farms of around thirty to thirty-five acres would be considered to be self-sufficient. In the hills of Wales farmers concentrated on sheep and cattle rearing; in the valleys there were dairy farms and some wheat growing.

As with the rest of Britain, farming practices began to change during the nineteenth and early twentieth centuries. Common land was enclosed, new equipment began to appear, there was more scientific stock breeding, new crops such as turnips were introduced, as were more resilient versions of existing crops, and there were changes in the relationship between landowners and tenants.

Wales also was affected by increased competition, but this brought with it the opportunities for new markets for farming products. There were faster ships and there were railways, new fertilizers began to be introduced, crop rotation was routinely used and farmers were far more prepared to switch from one type of production to another, moving into livestock when corn prices fell.

There was a major agricultural depression in the 1870s, which saw an increase in grain prices. There was also considerable emigration: many people left more rural Wales to head for the south Wales valleys, where there was industry and guaranteed jobs, whilst others made for London, the United States, Australia or New Zealand.

During the course of the first few decades of the twentieth century the number of farms actually halved. The greatest period of change was 1890 to 1918. The focus had been on producing more for home consumption rather than export. Huge estates began selling off their land and many farmers became more independent. In Cardiganshire farmer's cooperatives were set up in 1903.

Welsh lamb is, perhaps, the first agricultural product that springs to mind when considering farming in Wales. Indeed the breeds that were developed in Wales were specifically selected for their abilities to cope with harsh mountainous conditions. The Welsh mountain sheep, which is in fact a range of related types of sheep mainly with different colour variations, has no wool on the face or legs. Their tails are normally left undocked. The Welsh mountain breed has shown that it can cope with the harsh environment of its natural habitat. It is happy with high rainfall and even freezing winter temperatures. Understandably, breeders were keen to ensure that they were hardy, that the ewes were good mothers and that the lambs were strong and would survive.

Wales is around a quarter of the size of Scotland and has huge areas of mountainous terrain. It also has a highly productive coastline and the main industries have tended to be in the south of Wales, particularly around the coal mining areas. At the turn of the nineteenth century Welsh would probably have been spoken far more widely than even fifty years later, possibly even more than it is today, with the various programmes designed to encourage the Welsh to speak their native language.

Understandably, Welsh cuisine is highly influenced by the immediate availability of farming produce. With around eighty per cent of Wales being used for agricultural purposes the vast majority of it remains grass pasture or rough grazing for sheep and for cows. It is only in areas such as Carmarthenshire and Pembrokeshire that there are any large beef and dairy herds.

Welsh laver bread contains boiled, minced or pureed seaweed and is traditionally eaten with bacon and cockles, all products of the extensive coastline. Although something of a stereotype, lamb stews known as cawl, effectively a soup or broth, are still popular, as of course is the leek (cawl cennin is leek soup).

This is a photograph believed to have been taken in 1897, showing hand-milking. Milking cows twice a day was a chore that was usually reserved for women. In fact, hand-milking was standard practice until at least the 1930s. This photograph was taken in a remote farm at Felin Newydd in Carmarthenshire in South Wales. Open pails became buckets with lids after the Milk Regulations came into force in 1885. Milk had to be stored and cooled in churns and taken away every day for collection. On isolated farms such as this one, with impassable farm roads in winter, this would have to be done by sledge.

A farmer and his son can be seen here in Llanberis, near Mount Snowdon. This village, now in Gwynedd, takes its name from the Welsh Saint Peris. Originally the village was well known for its slate quarrying and it is in an area that has always been noted for its rugged beauty. Snowdon itself is the highest mountain in Wales and is now part of Snowdonia, the third national park to be created in Britain, established in 1951. The area retains many of its farming traditions and within the park boundaries there are still 26,000 people. Although Snowdonia now attracts six million visitors each year and the majority of the land is open pasture or mountainous, there is still significant agricultural activity. It is believed that people have been using the land around Snowdonia for at least 5,000 years. In many respects this area of Wales reflects the type of farming that has been going on in the country for centuries. Because of its historical interest and degree of preservation it is still possible to see the fields and their shapes, the old field boundaries and the farm buildings themselves. Agriculture still plays an important role in the area and for decades from the 1700s farming families would move their animals from the coast or the valleys to graze their stock on the sides of the mountains, where they could eat the fresh summer grass.

A somewhat stylized, but traditional, Welsh rural family at the beginning of the twentieth century is shown in this photograph. Just over twenty per cent of the modern-day Welsh population speaks Welsh. The language itself is derived from an ancient Celtic language that emerged in the sixth century. There has always been considerable dispute and debate about the teaching of Welsh and after a considerable struggle Welsh is now an integral part of the school curriculum. At the end of the nineteenth century the Welsh language was often referred to as Cymric, which simply meant Welsh, or relating to Cymry, which was a description of the Celtic language of Wales.

Two Welsh women are raking up hay from a small field in this photograph, with Snowdon in the far distance. Cattle were once very important within the Snowdonia area of Wales. From the late Middle Ages until the nineteenth century cattle would be driven to markets as far away as London, the Midlands or even East Anglia. It would take the drovers some two months to make it to London. Across Snowdonia there are still existing drovers' roads. In this photograph the two women are collecting hay that can be used for winter fodder. Traditionally, the summer farms and fields would be known as hafod and the farms and fields in the lower areas, which would be used in the autumn and winter, were called hendre. In order to preserve the way in which the fragile mountain habitat appears and serves the local wildlife, sheep are now not allowed to graze on the high mountains during the winter months. This also ensures that there is enough young grass in springtime, when the lambs are born.

Another shepherd is looking after a small flock in the Welsh Highlands in this photograph. From the Middle Ages the meat of the Welsh mountain sheep was prized for its excellence. Some areas specialized in breeding black-fleeced sheep. The rams had horns and the females were polled (hornless). Welsh mountain ewes on average produce two lambs each. The Welsh mountain sheep is, of course, not the only type: there is the Torddu, or badger-faced Welsh mountain sheep. They have distinctive black stripes above their eyes and black bellies. The Balwen is another rare breed that is perfect for the harsh conditions in Wales. It can be black, dark brown or dark grey. The black-faced but white-fleeced Llanwenog is also highly prized. The Lleyn is a lowland sheep and the ewes have a strong mothering instinct. The sheep in this photograph are likely to be standard Welsh mountain hill sheep. The majority would have white or tan faces. They have close-textured fleeces and an average ewe could weigh as much as forty kilos.

A photograph that is believed to have been taken at Nant Peris in the Snowdon range can be seen here, showing a farmer's wife milking a cow manually. Nant Peris was a small hamlet that is now better known as Llanberis – the original village became known as Old Llanberis. The church there is dedicated to Saint Peris, who was believed to have died in Nant Peris. He drank at the well where the church is now located every day. The valley itself, in the sixteenth century, was known as Nant y Mynach, or the Monk's Valley. Crowds of people would come to the well, where it was thought that if they drank the water after making an offering they would be cured of any diseases. The well itself had two trout swimming around in it and their appearance was supposed to signify that the person drinking the water would be cured.

A shepherd is attracting the attention of his sheep way up on a peak in Snowdonia here. Below the mountains are lakes and fertile valleys. There are far fewer sheep in the Snowdonia area than in the past. There is constant concern about getting the balance right so that the fragile habitats are not destroyed. Some areas have been fenced off for many years, in order to protect endangered plant life and small trees. Historically, shepherds were as aware of the environment as they are today and by moving their flocks on before they had overgrazed areas they helped to ensure that many of the plants had a chance to recover. Nonetheless, many areas were overgrazed and certain plant species did disappear, which had a knock-on effect on the wildlife.

This photograph shows the backbreaking process of loading wheat or oats on to a horse-drawn cart in the field. This process would have been unchanged for generations until the introduction of the tractor and more mechanized methods. Farmers have traditionally grown wheat for a number of different reasons in Britain. Hard wheat that is high in protein and has starchy gluten is used for bread making. Soft wheat with low protein and weaker gluten is used for general flour uses and lower-quality wheat is used for animal feed.

Chapter Five

Farming in the 1930s

This collection of fascinating photographs, courtesy of James Payne, has never been published before. They primarily focus on haymaking and harvesting oats. It is believed that the photographs were taken either in Kent or Surrey in the early 1930s.

Haymaking is a long-established means of conserving grass for feeding livestock through the winter. It has certainly been an important part of the farming calendar in Britain for the last 6,000 years. Farmers have long since known that successful haymaking relies on the crop of grass being thoroughly dried before it is baled or stored.

The normal procedure is to mow the grass. Obviously in modern times this is done with specialized machines, but in the past it would have been carried out either by hand, then by the use of horse-drawn implements and finally tractors. The process usually begins around June. Ideally, the mowing should happen before the flowering, but in many cases the cutting takes places during the flowering (which then produces a great deal of pollen and hay fever).

Cutting needs to happen when the weather is fine and the farmer has the reasonable expectation of several days of dry conditions. Grass cut when wet produces poorer-quality hay that the animals find less palatable. The grass then has to dry in the sun. To encourage this, the grass is either stood up in stooks or simply raked over to help remove the moisture by exposing the lower layers to the sun.

Effectively, haymaking transforms green, perishable forage into a product that can be safely stored and easily transported without danger of spoilage. It also keeps losses of nutrients to a minimum: this means reducing its moisture content from between seventy to ninety per cent to twenty to twenty-five per cent or less.

A horse-drawn binder or reaper makes short work of the oat harvest. This type of machine represented the next stage in development with the addition to the machine of a seat for carrying the raker. The machines built in 1831 required that the raker walk by the side of the machine. In 1845 McCormick added a seat, patented in 1847. This seat enabled him, while riding, to rake the grain from the platform and deposit it in gavels on the ground. This type of reaper, patented in 1847, is based on one taken by Cyrus H McCormick to the Great Exhibition, held in London in 1851. It was said of the machine at the time: 'The McCormick reaper is the most valuable article contributed to this exposition, and for its originality and value and perfect work in the field it is awarded the council medal.'

The field is nearly finished as the last two wagons are filled up by hand. This type of hay gathering would have been the traditional form, whereby hay was mown, turned and then carted. As mechanization came in, chopped hay would prove to be less bulky. Mechanized baled hay, which had been hand baled or baled by stationary machines, would not be introduced until the 1950s. Nowadays, round bales are the most popular, as the shape sheds rain and resists water better than traditional rectangular bales.

This is a hay loader in action near London in the 1930s. For generations hand-trussed hay was used in manual haymaking. The hay loader would elevate the hay straight on to the wagon. By around 1879 around 100,000 of these machines had been made. The hay loader ended the manual pitchforking of the hay on to the wagons. The horses would be led down the rows of hay with the hay loader and a wagon behind them. The loader would be able to pick the hay off the ground and lift it into the air and deposit it on to the wagon bed.

This is the first of four photographs that illustrate various stages of the hay harvesting. This photograph shows a very basic mowing machine. Deciding whether or not to mow on any given morning was always a hotly discussed topic. In the uncertain British climate, the farmer had to work out whether there would be a spell of good weather sufficient to allow the making of hay. At the very least, he had to be sure that he could get in into large windrows (rows of cut hay). Experience always counted, as the time the hay took to cure (dry) meant that he needed at least three days.

Practical horse-drawn mowers had been developed as early as the mid-1840s. Certainly by the 1880s, the designs were pretty standardized. They were straightforward in design, consisting of a two-wheel cart with an attached cutter bar. Gears and a shaft transformed the rotary action of the cart axle into a reciprocating action, which in turn drove the knife (serrated) in and out of slotted guards, which were attached to the cutter bar.

Each manufacturer had different gear systems. The most common were the two-horse mowers, which had five- to seven-foot cutter bars. There were also one-horse mowers that had three- or four-foot cutting bars. The mowers themselves needed constant attention. The knife on the cutter bar had to be very sharp: if the knife dulled then the mower would clog up and eventually stop. It would also mean that the horse or horses were being overworked. As a result, the farmer would, more often than not, sharpen the knife every day. Each part of the knife had to be ground by hand with either a file or a specially shaped grindstone.

The operator of the mower needed to be very vigilant. Note the way in which the operator in the photograph has cocked his head and is looking down at the cutter bar. He needed to make sure that the cutter bar did not strike a hard object, such as a stone. Even the smallest stone could cause enormous damage to the cutter. The horses also needed to be well trained for the job: they were often very jittery because of the sound of the whirring just behind them. Mowing had to done as soon as the morning dew had dried, as if the grass was still wet, then the knife would jam. Farmers leapt on warm dry days to get the first stage of the harvest done.

This is a hay tedder at work. It was used after the grass had been cut and before the grass was windrowed. The hay tedder came into operation in the second half of the nineteenth century, used to 'stir' up the cut grass and spread it so that it dried more evenly on the ground. The development of this piece of machinery meant that it saved enormous amounts of time and money. One man and a horse could easily do the same amount of work as up to fifteen farm labourers. It also meant that the cut grass could be turned on the same day, even if it had got wet or been trampled.

This photograph shows a horse-drawn binder at work. The oat or corn crop was cut with a binder, which was drawn by two or three horses. As the sheaves left the binder they were often hand lifted from the ground and put together in heaps of eight or ten to form stooks. This allowed the air to circulate around the sheaves and dry them. The sheaves were automatically tied with string by the binder.

After the grass was cut, the farmer relied on the sun and wind to dry it. It was then raked into rows using a horse-drawn rake. It would then be separated and forked into haycocks, which were about a yard high. These were small enough for one man to fork the haycock on to the loading cart.

This is the last load of hay ready to go to the hayrick. Corn and haystacks had to be constructed carefully so that they would not collapse. The heads of the corn went into the middle of the stack, making a circle as the sheaves were forked from the cart to the person making the stack.

This is a hayrick near Hertingfordbury, Hertford. Once the rick was finished it was thatched with wheat or rye straw. The straw was then secured to the rick using twine that was tied to short pegs hammered into the hay. Depending on the area of the country, ricks had different shapes and styles. In the Chilterns, for example, they would be oblong with gable (vertical) ends. Bales of hay could be cut from the rick as needed using a hay knife. This was called hay trussing.

Preparing the soil for the next year's harvest, using a Fordson tractor, is seen in this photograph. The tractor was not without its problems as it was quite lightweight – the wheels slipped in some conditions and it also suffered from rearing backwards if the plough hit a major obstruction.

This is the Post Mill on Reigate Heath in Surrey. It was built around 1765, although a mill was marked on maps dated 1753 and 1762. The mill was last worked by wind in April 1862. In the 1960s it was fully restored and is now believed to be the only consecrated chapel in a mill in the world.

This is a Case tractor undergoing trials. Case was another US manufacturer that began by developing reaper machines back in the 1830s. They made traction engines and later cars. Over 30,000 of their steam engines were also manufactured.

This early tractor is cutting the crop. The first powered farm implements were portable engines, essentially steam engines on wheels that drove machinery using a belt. Traction engines came in around 1850. The first recorded tractor sale was in 1897 but the first commercially successful design dates from 1902. The first four-wheel design came in around 1908 and by the 1920s tractors with internal combustion engines had become the norm.

A more traditional way of harvesting, using 2hp, can be seen here, rather than the early tractors, which had around 18hp.

Problems with the harvest: a boy is watching his father sort out a blockage here. For the first time tractors gave farmers the opportunity to harvest effectively for themselves rather than have to bring in casual labourers or work together with neighbouring farmers to harvest one another's fields systematically. Although the capital investment in these tractors was huge they would pay back the investment after just a handful of harvests.

An International tractor is shown here, which was another American-built machine. They entered the market with gas-powered tractors but in 1932 introduced their first diesel engine. By 1935 the WD40 became the world's first diesel tractor on wheels. Throughout the early period they were in direct competition with the Fordson, which was far more widely used.

This is a hay loader in action. The hay loader was about six or seven feet wide and was basically an angled, flat wood surface. About a foot above this surface were five or six narrow boards that travelled the length of the loader. Each had a number of spikes pointing toward the loader's bed. The boards were mounted at the bottom of the loader on a shaft, just like the crankshaft of an engine.

Here, a man feeds the loader with a pitchfork whilst two others work on the haystack. The loader itself was 'ground-driven'. There was a chain powered by a large sprocket inside one wheel of the loader. This provided the power to turn the shaft. The movement of the boards pushed the hay up the bed of the loader, which had spikes. The board would then lift away from the hay and move down to impale a new load of hay. At any one time, half of the spikes would be in a position to hold the hay and prevent it from slipping down the ladder.

This photograph shows the next logical mechanical development of the process, with the introduction of the tractor as part of it. The mown grass was known as swath and a windrow was made by raking one or more swaths together (this could also be formed by a mower or a windrower). After the swath was dried, further drying might be needed in the bale, stack, barn or field. Farmers made a judgement about the harvesting based on the amount of drying time that would be needed for the swaths and windrows. The actual quality of the hay was judged by the eye, or by handling (usually twisting a bunch of hay). It needed to feel crisp and look bright. If it looked bleached then this indicated a badly weathered crop.

One thatched haystack completed as attention turns to the second stack, using the hay loader. Stacking was a very skilled job and the quality and safety of the harvest depended heavily on the skill of the builder, especially in countries such as Britain, where there was always the risk of moderate to high rainfall. Ideally, when building the stack, the centre needed to be tightly packed, otherwise the stack would sag as it settled and rain would penetrate the hay. The sides would normally slope outwards slightly to the eaves and the head would rise gently to a point or ridge.

Two horses pull a plough whilst the ploughman struggles to keep his line. Ploughing developed its own language over the years. A backstroike meant ploughing the same area twice; cooms were the ridges formed between horse-tracks and wheel-ruts; hidlan(d)s were the outside edges of the field; ollan(d)s were old pastures that had been ploughed for crops; pan was the opposite to ploughing, that is, when the ground had been flattened down by heavy rains; and a rig was the space between the furrows.

A farmer and wife hold a pair of working horses. Ideal farm horses needed to be fairly tall and extremely muscular. As a result, heavy horses such as the Shire, the Suffolk and the Clydesdale were the real favourites as far as farmers were concerned.

Local farmers viewing the latest in ploughing technology that the 1930s could offer. A local farmer would host events showing new products and machinery at work. This is still very much a practice in use and an ideal way to market new ideas as part of a social event.

This appears to have been taken at a ploughing championship. The British National Ploughing Championships still run today with around sixty events taking place each year. Classes include conventional, reversible and horse-drawn ploughs, as well as those for vintage tractors and ploughs.

Building some magnificent haystacks alongside a farm track. Sometimes, ropes or nets were used to protect the stacks against wind. Thatching with coarse grass or straw was common in areas of the country where rain was expected during storage. The farmers might hold back from this part of the process in order to allow the dissipation of heat and moisture following stacking.

Two men shock up the oats (form stacks) during the harvest. Oats were commonly grown as feed for ruminants and horses. Since the more widespread introduction of the tractor in the 1950s, oats declined in importance and barley offered better value to the farmer as animal feed. The harvesting of oats was also difficult as the grain and ear fell off very easily if it had become overripe. Oats are still grown in the north of England and in Scotland, both for human and animal consumption.

This is the more modern method of sweeping the hay, fully automated and driven by a tractor rather than horse power. The purchase of hay loaders such as these meant an end to the manual pitchforking of the hay on to the wagons in the hayfield. The tractor could be driven down the windrows of hay with the hay loader and a wagon behind it. The loader would be able to pick the hay off the ground and lift it into the air and dump it on to the wagon bed.

A side-delivery rake in action – note the iron wheels. Although rubber wheels were commonplace on motorcars by the 1930s, the use of iron wheels persisted. This type of rake left a windrow of hay to the left and beside the rake. Before balers came into use, the trip rake was used. It operated rather like a normal garden rake and when it was full it would trip, lifting the teeth and dumping the hay. The side-delivery rake used a series of rotating bars (ground driven by capturing the power of the wheels), which meant that the rake always worked in time with the ground speed. It had large wheels in front and smaller wheels at the back and worked well when pulled by a horse and on relatively flat ground. On a hillside the rake would tend to swing downhill.

A photograph showing the side rake in action. Often the purchase of a horse-drawn side-delivery rake would mean retiring the dump rake to permanent storage. The side rake was of a completely different design, with four pipes to which many small teeth were attached. The pipes were attached to and spun around a central shaft; the movement moved the hay into windrows rather than simply being caught up in the teeth of the dump rake.

This photograph shows the far more traditional way of hay sweeping by using two horses working in unison. Horse-drawn sweeps were developed to push several sections of windrow together to make collection easier. Some would have a tumbling action so that they discharged more easily.

Hay-sweeps were large, stout, coarse rakes, with teeth projecting both ways. A horse was attached to each end (sometimes with a boy on each horse). A horse passed along each side of the windrow, drawing a rake after the pair, scooping up the hay as they went. When around 500 pounds had been collected they would draw it to the stack or barn and then the horses would turn about. This would cause the gates to make a half circle, draw the teeth backwards from the heap of hay and then empty for another load.

Thatching ricks: a lost art. The thatch covering on a stack was only a temporary form of covering, unlike the thatch on a house that could be expected to last for many years. Thatching experts were in great demand: the number of thatched buildings in England decreased from around 1,000,000 in 1800 to 35,000 by 1960. In Britain today there are around 24,000 listed thatched buildings and around five per cent of new-build housing has a thatched roof.

There are obviously problems with the boiler during threshing. The threshing machine (or thresher) was invented by Scottish mechanical engineer Andrew Meikle in around 1784 in order to separate the grain from the stalks and husks. Some of the steam-driven threshers have been preserved and can be seen at steam festivals and traction engine rallies. One notable event is the Great Dorset Steam Fair, where four or more of the steam threshers are in operation at any one time.

The hard work is done, the harvest is in and the farm returns to normality. Harvest is derived from an Anglo-Saxon word 'haerfest', which meant 'autumn'. It later came to refer to the season for reaping and gathering grain and other crops. The full moon nearest the autumnal equinox is known as the harvest moon. In ancient traditions harvest festivals were held on or near the Sunday of the harvest moon, usually in September.

Threshing oats in the barn with a tractor providing the belt-driven power. Often the largest building in the farmyards was the barn used for the process and storage of corn. The entrance needed to be big enough to admit a high-loaded wagon. The load was dumped into bays either side of the door. The bays were designed to prevent the crops going mouldy, so they needed to be well ventilated. The floor surfaces were either stone, brick or earth. The threshing floors in two-storey barns would be thick oak planks. For ground-floor threshing brick, earth or clay, stone flags and cobbles were used.

The threshing continues in the field, even though the boiler has stopped working. The development of threshing machines with mechanized feeding, threshing, winnowing and straw-stacking devices meant enormous power was needed. The steam engine would provide continuous power and increased output. As long as the steam engine was kept at operating pressure there was an unending supply of energy to run the machinery.

This is a photograph showing the wheat being threshed direct from the field. The 1881 *Household Cyclopedia* said of Meikle's machine:

> Since the invention of this machine, Mr. Meikle and others have progressively introduced a variety of improvements, all tending to simplify the labour, and to augment the quantity of the work performed. When first erected, though the grain was equally well separated from the straw, yet as the whole of the straw, chaff, and grain, was indiscriminately thrown into a confused heap, the work could only with propriety be considered as half executed. By the addition of rakes, or shakers, and two pairs of fanners, all driven by the same machinery, the different processes of thrashing, shaking, and winnowing are now all at once performed, and the grain immediately prepared for the public market. When it is added, that the quantity of grain gained from the superior powers of the machine is fully equal to a twentieth part of the crop, and that, in some cases, the expense of thrashing and cleaning the grain is considerably less than what was formerly paid for cleaning it alone, the immense saving arising from the invention will at once be seen.

This is the last photograph showing the wheat being threshed direct from the field. The 1881 *Household Cyclopedia* went on to say:

> The expense of horse labour, from the increased value of the animal and the charge of his keeping, being an object of great importance, it is recommended that, upon all sizable farms, that is to say, where two hundred acres or upwards, of grain are sown, the machine should be worked by wind, unless where local circumstances afford the convenience of water. Where coals are plenty and cheap, steam may be advantageously used for working the machine.

Chapter Six

1930s Rural Britain

Britain's cities have changed beyond recognition, even in the past few generations. Changes to the rural countryside have been more gradual. Yet we have certainly seen a creeping urbanization. In fact a report by the Campaign to Protect Rural England in 2005 suggested that creeping urbanization and development would lead to the virtual destruction of the traditional English countryside by 2035. It blamed the housing shortage, industries setting up in the countryside, increased road building and traffic and airport construction. Certainly in living memory there has been a steady decline in traditional farming and with it the whole fabric of rural England.

We may still have quaint villages, chocolate box cottages and tiny communities nestled amongst the folds of the British countryside. However, this collection of photographs, which were taken in the 1930s, primarily in the south of Britain, show just how much even traditional and relatively isolated rural villages have changed. These small, self-sufficient communities were not only the homes of the farm workers and those that relied on the farming industry for their income, but also the whole infrastructure of the local community: the doctors, storekeepers and, of course, the public house.

Many of these villages have now been transformed into commuter communities. They lack the infrastructure and services that inhabitants only seventy or eighty years before relied upon for their entire existence. Over the decades we have seen countless post offices, rural shops and public houses close down. There have been tremendous changes in shopping and social habits.

Mercifully many of the buildings that did not succumb to bulldozers before conservation of our built heritage became important still survive to this day. This is a fascinating collection of photographs. Unfortunately we do not know the photographer, but it is clear that whoever this was he or she had a keen eye for what was even then becoming a rare rural sight.

Governments were already aware of the problems facing rural Britain between the wars. In 1926 the pioneering planner Sir Patrick Abercrombie wrote *The*

Preservation of Rural England. It called for a national committee to be formed in order to preserve the countryside. It was created in December that year, as the Council for the Preservation of Rural Britain. Abercrombie became its first honorary secretary.

By the mid-1920s road building was cutting swathes through the countryside. But there was a call to set up national parks and to protect rural Britain. By the early 1930s, as electricity cables were being put in position across the country, steps were taken to try and minimize the visual impact on the countryside. It was in 1933 that Raymond Unwin proposed what he called the Green Girdle for London. This was, of course, to become what we now know as the Green Belt.

Unwin was an influential urban planner. He was a Yorkshireman who had moved to Oxford. By the 1930s he was president of the Royal Institute of British Architects and had become highly influential. He proposed that there be an open belt of protected countryside all around London. This was to prove to be something of a blueprint, which to some extent continues to this day to protect rural areas from urban sprawl.

The countryside would prove to be incredibly important for the war effort and in fact the wartime aims of the Campaign for the Protection of Rural Britain were 'The conservation and development of our agricultural resources and the improvement of the social environment of the rural population.'

Directly after the Second World War the Town and Country Planning Act (1947) came into force. This now meant that there would be public scrutiny of all planning enquiries in the countryside. The first ten national parks were designated in 1949 but by the 1950s rural Britain was coming under enormous pressure and the green belts around Britain's largest cities and towns were already being encroached upon, not only by housing but by industry and roads.

Over the next few decades there were battles to save parts of rural Britain. In 1963 the Berkshire Downs were saved from the ravages of the construction of the proposed M4. Important wetlands in West Sussex were protected from drainage in the late 1970s. In the 1980s proposed extensions to the M40 only served to highlight the threat to unspoilt countryside. Rural Britain is no safer now than it was through these decades. Many small rural towns suffer from a lack of development, traffic congestion and a declining high street.

In the 1930s village and market town life was entirely bound up with farming. The majority of the men worked on the land as farm or estate workers. Women came out to the farms to help, particularly during spring time when they weeded crops and during the harvests in the summer and autumn. Tradesmen toured the countryside in their vans, selling groceries and fuel. In many areas of rural Britain there was no electricity supplied until the late 1930s. Paraffin was used for heating, lighting and cooking. The community revolved around the churches, schools and public houses. The pace of life was slow. The village cottages were somewhat dilapidated. For many of the villages it was a life apart, a self-contained existence.

We can see what life may have been like in the 1930s by visiting traditional farms like Tatton Park in Cheshire, or Treberfedd in mid-west Wales, near Cardigan Bay. Undoubtedly it was a hard life and during the 1930s farming was going through a depression. It was still very labour intensive and most of the farm work was still carried out by human and animal muscle. Prices for farming products had fallen by as much as sixty per cent, unemployment was high and there were few other jobs. This was to mark another cycle of exodus from rural to urban lives for many families.

This is the first of five photographs that were taken in Aldbury during the 1930s. This row of terraced, thatched cottages is just to the north of the village pond on Stocks Road, which leads to Ivinghoe. Aldbury was chosen as one of the locations for *Midsomer Murders*. The village is in Hertfordshire and sits below Aldbury Common, on Ash Ridge, which forms part of the Chiltern Hills. The village is around two miles from Tring. Most of the older buildings are around the village green and pond, where some old stocks and a whipping post can still be seen. It is believed they were last used in 1835 to punish a local drunk.

This is a panoramic view of Aldbury from Moneybury Hill. This is the highest spot in the parish. Clearly visible is the church that stands at the north-west corner of the village, on the Tring Road. This is the Church of St John the Baptist and it probably dates back to the thirteenth century, although there were additions to it for at least another 200 years. Standing on Moneybury Hill is the Bridgewater Monument, which was built in 1832. It commemorates the Duke of Bridgewater, who designed Britain's canal network.

This photograph shows the pond at the centre of Aldbury in the late afternoon. The parish of Aldbury, according to an early twentieth-century history of Hertfordshire, comprised of just over 2,000 acres of land. Around 600 acres of it was arable land, 200 acres was grassland and nearly 280 acres was woodland. The crops that were grown in the area were mainly wheat, barley, oats and roots. The village itself, as can be seen in this photograph, lies in a valley with chalk hills either side. The hills are covered with beech and fir trees. Some of the buildings that can be seen to the left of the photograph are believed to date back to the first half of the seventeenth century. This particular house has a tiled roof, but the others are thatched, which is relatively rare for this area of Hertfordshire.

Although this photograph is not identified, it may well have been taken outside of The Valiant Trooper public house, located in the village of Aldbury. The Valiant Trooper consists of a number of seventeenth-century cottages. They have been ale houses for several centuries. Researchers have discovered that in 1752 the ale house was known as The Royal Oak and was owned by a John Barnes. Inside the pub and carved in the brickwork are the initials ID and the date 1769. This was presumably carved by the next owner, Isaac Dell Master. The public house became The Trooper ale house in 1803 and rumour has it that the Duke of Wellington was a visitor and he discussed tactics with his officers and men. It became The Valiant Trooper in 1878. At this stage in the 1930s the public house would not have resembled anything like a modern-day pub. It would have been more of a taproom, and there would be no bar area as such but simply a series of rooms with benches and tables, with one of the rooms set aside for the barrels. Several examples of these original taproom pubs still exist, including one known to the authors in Laxfield, Suffolk. The Laxfield Low House is in fact called The King's Head, and it is largely unchanged since Victorian times. The Low House, on Gorams Mill Lane, is now an Adnams public house, still retaining its original taproom.

An inhabitant of one of the cottages on Stocks Road can be seen in this photograph. The village itself is recorded in the Domesday Book, where a manor is mentioned. By 1271 there were sixty-eight households in the village. Today, including the houses at Tring station, there are around 1,000 inhabitants. The village of Aldbury lies just below the National Trust's Ashridge Estate. There is a vast parkland that was originally a deer park and around 2,000 acres of mature woodland. Close by is one of England's oldest surviving post mills, which was an early form of windmill. It dates back to 1627 and was restored around fifty years ago. This is situated on Pitstone Hill.

This is Benington, some five miles to the south-east of Stevenage in Hertfordshire. Visible are the cottages in Town Lane. Benington is believed to date back to Saxon times. It was a fortified site used by the Mercian kings. Following the Norman conquest of Britain the manor was given to Peter de Valoignes and there are still remains of a motte and bailey castle. From records dated just prior to the First World War Benington parish was just over 3,000 acres. The population has barely doubled since the 1840s, when it stood at 605.

A close up of the cottages on Town Lane are featured in this photograph. The village has a strong farming history, with virtually all of the surrounding countryside set aside for agricultural work. One of the two public houses in the village is The Bell, which dates back to the fifteenth century, and this building is also situated on Town Lane. The other pub is The Lordship Arms, on Whempstead Road. Photographs still exist of The George and Dragon public house, which was still listed as a retailer of beer in 1912. Postcards also exist of this establishment, which were actually issued by the post office in Benington.

This is a photograph of the post office in Benington, also on Town Lane. One of the key events of the farming week since the fourteenth century was a weekly market in Benington. In fact a market and fair continued on the same spot every year for 600 years. It was primarily a centre for trading and with merchants attending the market town's event there were sideshows and stalls. The fair still continues. A historian writing in the middle of the last century about Benington fair said:

> A more picturesque scene than Benington fair as it now exists cannot be imagined. The entertainment is not, it must be admitted, of a very intellectual nature, but it thoroughly comes up to the expectations and desires of those for whom it is designed and is certainly far less debasing than many of the exhibitions presented at London theatres.

The fair is held on or around the second Saturday in July. Originally it was given permission to be held on the feast of St Peter and St Paul, 29 June, but the fair was always held around 10 July.

This is a photograph of Hill House in Denham, Buckinghamshire. This was once the home of the actor Sir John Mills, CBE (1908–2005). Hill House is a late-seventeenth-century brick building with a triple gable front. The area around Denham, or, as it was originally called, Daneham, has a soil mixture of loam and gravel, with a chalky subsoil. As a result the chief crop is cereals. The whole area is heavily wooded. The village's name, which has had various spellings and is referred to in the Domesday Book as Deneham, means 'homestead in a valley'. Modern Denham is split up into several recognizable areas. The village itself is the original settlement. Denham Green is clustered around Denham station, New Denham is along the old Oxford Road and Higher Denham was built on the site of a First World War army training and transit camp. Sir John Mills is not the only notable resident associated with the village. Roger Moore, Mike Oldfield, Shane Richie, Richard Lindsay, Paul Daniels, Jess Conrad and Brian Connolly have all lived or still live in Denham.

We can see in this photograph Denham's Village Road. Two public houses can be seen in the photograph. These are still open and situated right next door to one another. The public houses are The Green Man and The Swan, which is a Georgian double-fronted building. Village Road largely comprises red-brick houses with giant wisterias. This has made them a popular location for many British films and television programmes. In fact, Denham studios operated near the village from 1936 to 1952. Some of the notable films made there were *Goodbye Mr Chips*, Noel Coward's *In Which We Serve* and Laurence Olivier's *Henry V* and *Hamlet*. Part of the shooting for David Lean's *Brief Encounter* was also done in Denham.

This is another shot of houses on Village Road in Denham. These houses have barely changed since they were built. The majority of the houses and cottages in the village date back to the sixteenth and seventeenth centuries. Some of the oldest houses are timber and brick cottages and are located to the north-west of the church. The old school dates back to 1721.

The bridge over the River Misbourne in Denham can be seen here. The Misbourne is a tributary of the Colne River, where traces of ancient fishponds still exist. There was originally a large fishery there. The Grand Junction Canal also passes through Denham. The Grand Junction Canal was once a busy route and vital for farmers. There was also a passenger boat service. The Paddington stretch of the canal opened in 1801, which terminated in a 400-yard-long basin, some 30 yards wide. Here there were wharves, a hay and straw market, pens for livestock and warehousing. This illustrates the point that farmers all along the canal, including Denham's, relied on this transport network to transport their produce and livestock into the capital. On 1 January 1929 the Grand Junction Canal became part of the newly formed Grand Union Canal Company. The 137-mile canal stretched from Salford Bridge Junction in Birmingham to Brentford and it became the longest canal in the country.

In this photograph we can see the first of a series of shots that were taken presumably in the Radlett area of Hertfordshire. The photographs illustrate the process of road mending and improvement, much of which was carried out by hand. What is particularly perplexing about this sequence of photographs is the apparent locations of the two businesses involved in the operation. The contractors carrying out the manual work are Burrell & Co of Radlett. This business still exists as a builders' merchant and is still based on Watling Street. The second business, whose name can be seen on the steamroller, is Barnes Brothers of Southwick, Trowbridge, in Wiltshire. According to records dating back to 1927 Southwick had a shop, a garage, a post office and telephone call office, a blacksmith, a coal dealer, a house decorator, a stone mason and the steam haulage and removal contractors, Barnes Brothers. Some forty years before, in 1890, two related businesses, although unnamed, were a road contractor and a threshing machine proprietor. Certainly until the 1930s and beyond the main commercial interest around Southwick was farming. What is particularly odd about Burrell & Co and Barnes Brothers working together is the fact that even today the two businesses are around two and a half hours apart. This makes it incredibly difficult to pinpoint precisely where this road mending was taking place.

This is a close-up of the steam traction engine, which is crushing the tarmac surface prior to the workmen removing it. The front of the traction engine reveals it to be an Invicta steam traction engine. However, the name Burrell adds a level of confusion, as Burrells of Thetford in Norfolk have a long association of steam engines of many different types. Many of them were used for agricultural machinery and in fact over the course of the history of the works in Thetford some 4,000 engines were made. Burrells were the first to introduce practical heavy duty traction engines that could be used on roads. Their engines are also particularly well known for their showman's engines, many of which still survive to this day.

In the early nineteenth century Burrell Brothers were manufacturers of harrows and rakes, but they designed their own patent agricultural machinery. One such early invention, in 1803, was the 'improved drill for sowing crushed oatcake manure with wheat, turnips etc'. Burrell Brothers would go on to design threshing machines, ploughs to tow behind traction engines, a wide range of basic agricultural engines, indirect ploughing systems, compound engines, straw elevators, straw benches, gorse cutters and even engines for roundabouts, dodgems and fairground rides. The business finally closed completely in 1930. We can only speculate as to whether there is a connection between the original Burrell Brothers of Thetford and the road haulage business in Radlett.

Workmen are busily carting away the crushed road surface with wheelbarrows here. In the background a fire is already underway to heat up the tarmac for the relaying process. It is interesting to note that there was no need for traffic signals or signs to warn about work on the road. The steamroller would also have been used as a stone crusher for road mending. A fly wheel would be run off the steam engine on to the stone crusher, in order to provide the hard core to strengthen the road surface. A possible explanation as to why the Burrell steamroller found itself in Radlett in Hertfordshire is that many district councils bought up second-hand steamrollers in the 1920s and the 1930s. It was around this time that they became responsible for road mending. They would also need tar boilers.

A vintage Invicta steamroller can still be seen in operation at the Beamish Museum of Northern Life near the town of Stanley in County Durham. The Invicta brand was made by the British agricultural engine and steamroller manufacturer Aveling and Porter. Thomas Aveling and Richard Thomas Porter went into partnership in 1862, developing a steam engine three years later. The business was based near Rochester in Kent; however, by 1919 the company had joined with the Agricultural and General Engineers, but this went into receivership in 1932. The company rose again in 1933 to form Aveling-Barford. They continued to make steam and motor rollers. The business still trades under the name of Barford and their prancing horse logo is still in evidence. They now make site dumpers that carry bulk materials, often on building sites.

One of the workmen contemplates the backbreaking job of loading the tarmac. Professional road-making contractors would often have huts on wheels that were pulled by the road roller, so that the engine driver and the co-worker could use it as a base and somewhere to sleep, as they travelled all round the country, moving from job to job. Many of these huts were made by Aveling and Porter in Rochester and were a common sight up and down the country. Effectively, they were wooden, rectangular boxes, with two windows and a door. They would run on solid metal wheels. Some of these huts still exist.

Roads were always a major fact of life for residents around Radlett, as the main road is in fact Watling Street. This is an ancient trackway that originally linked Canterbury and St Albans. It was the Romans that first paved the route and street literally means 'a paved road'. Before the Romans arrived it was a broad, grassy trackway. It would have been even at that time an important route for farmers and merchants.

Radlett itself is believed to date back to around 1453. Until the nineteenth century it was a predominantly rural and farming community. There were just a handful of buildings and the whole area was dominated by estates owned by the Kendals.

Tipping out tarmac is seen here. Note that the tipping mechanism of the truck is hand operated.

It is believed that Radlett derives its name from the old English meaning 'a junction of roads'. Originally a small number of buildings grew up where the route from Aldenham to Shenley crossed Watling Street. A painting dating back to 1796 shows just a handful of farm buildings and a public house. There is considerable dispute as to which of the two public houses that are still at the crossroads was the original one. It is either The Red Lion or The Cross Keys, which is now known as The Railway.

Big changes were to come to the area in the middle of the nineteenth century, with the large estates still renting out parts of their land to tenant farmers. But the railway was extended southwards from Bedford and a station was built at Radlett. This generated an influx of new inhabitants to the area. The first shop in the village was set up as a cooperative society in 1883. The owner of the land, Charles Part, donated it for Radlett Garden and the village hall.

Radlett really came into its own towards the end of the nineteenth century, when large land owners began selling off plots near the village centre, allowing more cottages and shops to be built. In the run up to the First World War, although Radlett continued to develop, the bulk of the population was still employed on local farms and estates.

The workmen are shovelling tarmac from the wooden wheelbarrows on to the road surface in this photograph. A popular folk tale is that in 1901, on a road near Denby ironworks in Derbyshire, the county surveyor of Nottingham, Edgar Purnell Hooley, noticed that a barrel of tar had fallen out of a dray and burst open. Someone from the ironworks had covered up the sticky tar with waste slag from the nearby furnace. This was the first ever creation of tarmac. It had been created by accident. Hooley immediately realized its potential and patented Tarmac in 1902.

The alternative origins actually date back to John Loudon McAdam, who had invented a road construction method known as Macadamization. McAdam was a Scottish engineer and he came up with the idea of using crushed stone bound with gravel on a firm base of large stones. He also created camber, which ensured rainwater drained off the road. The original work was carried out for the Bristol Turnpike, in the early part of the nineteenth century. It was much later that hot laid tarred aggregate, or tar-sprayed chippings, would be used. In fact, tar-grouted macadam was being used before 1900. Hooley's patent involved mechanically mixing tar and aggregate prior to putting it down on the road surface and then compacting it with a steamroller.

The men are still hand-laying tar and aggregate into the dip in the road, which had been created by the steamroller as part of the road mending process. Each parish was obliged by law to repair its roads. The parishioners would elect a person to serve unpaid for a year as a highway surveyor. He could call upon able-bodied men to work on road repairs for up to six days a year without pay. This system began to disappear by the middle of the eighteenth century and instead of using unpaid labour work was often carried out by paid workmen. This meant that only minor drainage and pothole filling was being dealt with. Main trunk roads were kept in better condition than others and turnpike trusts were authorized by parliament to charge tolls on certain roads, the proceeds of which would be used for repair and maintenance.

By the twentieth century even the minor roads were subject to far more traffic, particularly around the cities, where the enormous numbers of market gardens would generate processions of huge vans laden with vegetables, fruit and flour heading towards the London markets. Petrol-driven lorries were becoming cheaper to buy, which meant that many roads had to be widened or otherwise improved. This may well be the reason for the work being carried out in this series of photographs. Many private businesses up and down the country owned their own fleets of steamrollers and would contract them out to local authorities.

One of the most popular engines used by local authorities was the so-called Convertible. It could be used as a steamroller for road mending in the summer months and as a straightforward traction engine for haulage in the winter. The steamrollers were still being used as late as the 1960s, when major parts of the M1 were made with their help.